BATTLES FOR WARGAMERS

The Roman Civil Wars 49-45 BC

by
Terence Wise

A BELLONA BOOK
Published by
MODEL & ALLIED PUBLICATIONS LTD.
13/15 BRIDGE STREET
HEMEL HEMPSTEAD HERTS

1974

INTRODUCTION

The aim of this series is neither to compete with more academic studies of military history nor with books which deal with how to play wargames, but simply to supply wargamers with concise, basic facts about sets of battles within one theatre or campaign, thus providing him—or her—with the raw material from which to reconstruct battles from history. Nor is it the intention of the author to lay down specific methods of refighting these battles, since different players will undoubtedly wish to refight the battles from different stages of the action, but it is hoped that from the details listed players will be able to draw all the information they need to reconstruct the battles to suit their own methods. A list of my main sources is provided for those who wish to unearth more information than the scope of these books provides room for, or for those who wish to obtain a better understanding of the over-all scene.

As much as possible I have avoided the major battles which are already very well documented and have sought to include some of the smaller battles, which I personally consider more readily adapted to the limits of the wargames table. Information on these less famous but still important battles is also much harder to find and I hope by including such battles I will save players many frustrating hours of research and at the same time perhaps encourage them to widen their knowledge of the art of war.

All the books in this series follow a general pattern; name, place, date; the aim; forces engaged, including all arms and services, forces in support and reinforcements if applicable; a description of the terrain, including man-made defences; any major 'outside' events which influenced the decisions of the commanders; an account of the course of the actual battle, since the mistakes and results of these battles provide lessons and ideas for players; a summing up of the battle, pointing to moves which decided the outcome; and a list of casualties.

Each account is accompanied by a map and a photograph of the battle as reconstructed for a wargame. Information about contours was especially hard to come by and rather than have only partial marking of heights I have divided the high ground into three levels at the most, and these can be represented on the table by a corresponding number of layers of whatever material is being used to simulate high ground. The maps are complemented by the photographs which show the scene as if from higher ground and illustrate more clearly some of the possible alternative moves which could have been made.

Finally, brief but condensed guides have been supplied for dress, equipment and organisation so as to provide a starting point for those who are venturing into this particular period for the first time.

I hope all wargamers will find fighting these battles as interesting as I myself found the studying and refighting of them.

Terence Wise

CONTENTS

	Page
Special Note	4
Sources	5
Map Key	7
Summary of the Opening Moves	8
First Spanish Campaign: Ilerda	9
First African Campaign	15
Dyrrachium	21
Pharsalus	27
Second African Campaign: Ruspina	33
Second Spanish Campaign: Munda	39
Organisation Guide	43
Dress and Equipment Guide	44

SPECIAL NOTE

Neither the strategy nor the tactics of the Civil Wars can be said to have been outstanding, except perhaps in the extensive use of fortifications and other engineering works, and in a large number of the major confrontations the opposing armies drew up in line of battle only to stand down at the end of the day without having done battle. Those confrontations which did result in battles are often dismissed in a few words by Roman writers, most contemporary accounts being more concerned with who said what when. This makes it difficult to describe in detail the actual fighting, such vague expressions as 'the fighting lasted five hours' being common. (Writing about battles of the Civil Wars in one small book is not made easier by the fact that the fighting took place in a number of countries.)

Since these two points are in direct contrast to the earlier book for this wargaming period, The Second Punic War, in which both Hannibal and Scipio Africanus revolutionised the art of war. I have attempted to compensate somewhat for this lack of detail by breaking from the usual format of the series and including several accounts of entire campaigns rather than single battles. This should enable players to fight an interesting map game with the skirmishes transferred to the board, and still leaves the major confrontation (the battle for the hillock at Ilerda for example) as a board game to be fought separately or as part of the campaign. Conversely, the lengthy and involved siege at Dyrrachium has been left out, but one major incident from the siege is covered in detail. Some battles more fully described by the historians might have been included instead of adopting this method but they come more under the heading of native rebellions than civil war between Caesar and Pompey.

SOURCES

Caesar, trans. Rex Warner.	The War Commentaries of Caesar. Mentor Books, 1960
Caesar, trans. A.G. Peskett.	The Civil Wars. (Loeb Classical Library) Heinemann, 1951
Caesar, trans. J.F. Mitchell.	The Civil War. (This book includes accounts of the African and Spanish Wars by other authors.) Penguin, 1967
Fuller, J.F.C.	Decisive Battles of the Western World Eyre & Spottiswoode, 1954
Plutarch, trans. Rex Warner.	The Fall of the Roman Republic. Penguin, 1958
Shuckburgh, E.S.	History of Rome. Macmillan, 1916

(Map detail of Utica was taken from Scipio Africanus in the 2nd Punic War by H.H. Scullard, Cambridge University Press, 1930, and The Story of the Nations: Carthage by A.J. Church, Fisher Unwin, 1888.)

ROADS

BANK AND DITCH FORTIFICATIONS

TOWN WALLS

FORTIFIED CAMPS

LESSER FORTS

STREAMS

RIVERS

BUILDINGS

AMPHITHEATRE

THEATRE

POMPEIAN INFANTRY

POMPEIAN CAVALRY

POMPEIAN LIGHT TROOPS

CAESAREAN INFANTRY

CAESAREAN CAVALRY

CAESAREAN LIGHT TROOPS

POMPEIAN MOVES

CAESAREAN MOVES

HIGH GROUND

WOODS

MARSH

COASTLINE

SUMMARY OF THE OPENING MOVES

Julius Caesar began the Civil Wars by crossing the Rubicon, a river which formed the boundary between Italy and Cisalpine Gaul, in January 49 B.C. and by the unexpected speed of his advance managed to occupy Umbria, Picenum and Tuscany with a minimum of fighting. He then hurried down the Adriatic coast towards Brindisi where Pompey and the Consuls, who had fled from Rome, were assembling a fleet to transport their small and disorganised army to Greece. Caesar was just too late to prevent this crossing, an act which might well have ended the war there and then, and, since he had no fleet with which to pursue, marched on Rome, which capitulated without a blow being struck. In a little over two months Caesar had gained control over the whole of Italy.

Pompey was strongest in Spain, where he had seven loyal legions, and Caesar therefore decided not to wait for a fleet to be built in order to cross to Greece but to attack the Spanish legions as soon as possible and thus remove the threat to his rear before confronting Pompey himself. Caesar was delayed by the resistance of Massilia (Marseilles) but he sent Gaius Fabius ahead with most of the legions to seize the passes through the Pyrenees. Fabius found Pompey's generals waiting for his at Ilerda, modern Lerida, on the Sicoris (Segre) River, a tributary of the Ebro.

FIRST SPANISH CAMPAIGN: ILERDA

May – July, 49 B.C.

CAESAREAN FORCES. Commanded by Gaius Fabius, later by Julius Caesar. Six legions, including the IX and XIV, totalling about 32,000 men, with 5,000 auxiliary infantry, 3,000 veteran cavalry from the Gallic Wars, 3,000 newly raised Gallic cavalry and 900 cavalry of Caesar's personal bodyguard. **Reinforcements:** 6,000 Gallic cavalry and Ruteni archers. Total, about 50,000 men.

POMPEIAN FORCES. Commanded by Lucius Afranius and Petreius. Five veteran legions of about 26,500 men with 80 cohorts of Celtiberian, Cantabrian and Lusitanian auxiliaries, about 38,500 men, with 5,000 auxiliary cavalry. Total, 70,000 men.

TERRAIN. Afranius and Petreius had decided to confront Caesar in the region of Ilerda because of the outstanding strategic advantages offered by the terrain there:- the Sicoris and Cinga Rivers were only 30 miles apart here and both rivers were impassable except via a stone bridge at Ilerda. The Pompeians had gathered a great store of corn and other supplies in the town and had access to the fertile countryside across the river. Any attacking force would therefore be isolated between the rivers in an area incapable of supporting a large army for more than a few days.

Fabius built his camp about two miles north of Ilerda and attempted to win over the local tribes in order to procure supplies. When all available supplies were exhausted, in a matter of a few days, he had two wooden bridges built, about four miles apart, and used these to forage on the other side of the river. Since the Pompeians were also foraging here there was considerable skirmishing between the cavalry and the foragers had to be protected by many cohorts.

One day two of Fabius' legions under Lucius Plancus crossed the river by the southern bridge to forage but were suddenly out off from their wagons and supporting cavalry by a rush of water which broke down the bridge. The Pompeians realised what had happened when they saw the wreckage in the river and Afranius at once led four legions and all the cavalry across his bridge. Plancus took up a position on some high ground, with his two legions back to back to avoid being taken in the rear by the enemy's cavalry. This enabled him to hold out against the heavy odds until Fabius had time to send another two legions across the northern bridge. The arrival of these legions signalled the end of the fighting and both sides then returned to their camps.

Two days after this fight Caesar arrived, escorted by his bodyguard. The bridge was almost rebuilt and he ordered it to be completed that night, then rode off to reconnoitre the surrounding countryside. The next morning, leaving six cohorts to guard the camp and bridge, he marched the army towards Ilerda and drew up his forces in three lines before the enemy's camp to offer battle. Afranius led his forces out but halted them halfway down the hill. Realising the Pompeians had no intention of abandoning their advantageous position, Caesar ordered his third line to dig a ditch 15 feet wide facing the enemy, the work being hidden by the first two ranks so that Afranius was unaware of it until Caesar led his legions behind the ditch for the night.

After waiting under arms all night Caesar's legions resumed their work at dawn, a legion digging each of the other three sides of the camp while the remaining three legions faced the enemy to ward off any attack. The Pompeians did in fact advance to the foot of their hill in an attempt to provoke a battle and halt the work but now Caesar refused to be drawn out and after a short while the Pompeians withdrew to their camp. A rampart was added to Caesar's new camp on the third day and the baggage and six cohorts from the old camp brought forward.

From this new position of strength Caesar saw at once that if he were to seize the small hillock between the enemy's camp and the town he could cut them off from their supplies in the town and from the vital bridge across the river. With this intention he led three legions from the camp and formed line of battle, then ordered the front line of the legion on his left to advance at the double and seize the hillock.

Afranius countered this by sending cohorts from his centre with orders to occupy the hillock first. This they succeeded in doing, mainly because the Spanish legions had adopted the native method of fighting whereby every man charged swiftly to seize an objective without regard to formation. The Caesareans attempted to dislodge their enemies but the hill side was so steep that they had to use their javelins to help them climb it and therefore had no chance to hurl their weapons. This disadvantage, combined with the dispersed form of fighting adopted by the Pompeians, led the advanced cohorts to fear that they might be surrounded on their exposed flanks and in the resulting confusion they were driven back to their legion, which in turn also began to retreat. Cavalry were sent forward to guard this retreat by wheeling between the forces to present their shield side to the enemy. Caesar then led the IX Legion to support this wing and succeeded in beating back the attack but the IXth, pursuing their retreating enemy too far, pushed on until they found themselves at the foot of the hill on which Ilerda stands. Here they were confronted by a gradual upward slope which fell away steeply on both sides, leaving only just enough room for three cohorts to form a line. No reinforcements could be sent to the flanks, nor could the cavalry be used to cover a withdrawal, although they were posted at each flank in case they could help later.

For five hours the IXth fought on this cramped front, exchanging missiles with an enemy who had the advantage of the high ground and whose missiles therefore fell with much greater effect. Afranius sent cohorts via the town to replace the men as they became weary and Caesar was forced to employ the same tactic. At the end of the five hours the IXth was exhausted, the men were out of missiles, and the line was beginning to falter. In a last desperate effort they drew their swords and charged the enemy line. breaking the line of one opposing cohort and forcing the others to give ground until their backs were against the walls of Ilerda. Some of the Pompeians became panic stricken and slipped away into the town. This momentary victory enabled the IXth to withdraw unopposed, screened by the cavalry which had somehow managed to struggle on to the slope. Casualties in this battle were Caesareans, 70 killed and over 600 wounded; Pompeians, over 200 killed. After the battle Afranius had extensive earthworks built on the hillock and thereafter occupied it permanently.

Two days after this battle a great storm caused floods which destroyed both of Fabius' bridges, cutting the Caesareans off from their forage area. The floods lasted for several days and this, combined with enemy cohorts now stationed on the opposite bank, prevented the Caesareans from rebuilding the bridges. Large stocks of supplies were in fact on their way from Italy and Gaul and one of these convoys, escorted by about 6,000 Gallic cavalry and Ruteni archers, now arrived at the Segre, only to find no way of crossing the river. Afranius heard of their arrival and that night took all his cavalry and three legions on a night attack, his cavalry taking the convoy completely by surprise. The Gallic horsemen rallied superbly and although greatly outnumbered held off the Pompeian cavalry until the legions arrived, by which time the remainder of the convoy had escaped to high ground. About 200 of the archers were lost and a few of the cavalry.

The food situation becoming desperate, Caesar had his men build boats of wood and hides and one night these were conveyed on wagons some 22 miles to the south of the camp. Parties of men then crossed the Segre in the boats, and occupied and fortified a hill on the opposite bank.

An entire legion was then put across the river and by dawn a bridge was being constructed from both banks of the river. The bridge took two days to complete, after which the supply convoys were able to reach the Caesareans and foraging parties could once more cross to the countryside to the east.

That same day Caesar also despatched a large part of his cavalry across the bridge and these, falling unexpectedly on foraging parties, captured a large number of the enemy. Several cohorts of targeteers were sent to rescue them but the cavalry split into two groups, one to defend the prisoners and the other to attack the oncoming forces. The attacking force succeeded in isolating and annihilating one cohort which had drawn too far ahead of the main body, then the cavalry returned to their camp. The construction of this bridge, and the success of the Caesarean cavalry, shook the morale of the Pompeians badly and they began foraging by night. Five local tribes went over to Caesar and promised to supply him with corn, and a cohort of Afranius' auxiliaries also defected.

In order to avoid sending foragers and the cavalry all the way south to the new bridge Caesar now ordered several trenches, each 30 feet wide, to be dug into the river banks above Ilerda in order to divert part of the Segre and so create a crossing place. While this work was being carried out Afranius realised his foraging and communications would be seriously threatened and therefore decided to move to Celtiberia, where he could expect to recruit large numbers of auxiliary infantry and cavalry, and continue the fight from there. He gathered large numbers of boats at Octogesa (Mequinenza) on the Ebro about 30 miles to the south and ordered a bridge to be built with the boats. At the same time he sent two legions across the stone bridge to build a camp on the other bank, fortifying it with a rampart 12 feet high.

As soon as Caesar learnt of these operations he had troops work 24 hours a day at excavating the trenches until it became possible for the cavalry to ford the now divided river. This in turn spurred Afranius to move quicker than planned and, leaving two cohorts of auxiliaries to guard Ilerda, he took the remainder of his army across the river to the new camp and at dawn the next day began to march towards his bridge of boats over the Ebro, Caesar sent his cavalry to delay this march and, according to Roman historians, at the insistence of his men led all but the weakest men in the legions across the fords shortly afterwards. Large numbers of pack animals were stationed in the river above and below the point of crossing, cavalry rescued any man swept off his feet by the breast high water, and the crossing was achieved without the loss of a single man.

Although this crossing had necessitated a six mile detour Caesar's legions caught up with the Pompeians before 1500 hours. Afranius drew up his legions on high ground, ready for battle, but Caesar was content to rest his men, following when Afranius tried to resume his march, stopping when he stopped, so that Afranius was forced to camp much earlier than expected, only five miles short of the mountain passes which would give him safety.

The next day there was stalemate and both sides remained in their camps but before dawn the following day Caesar led his army by a most difficult route to get between Afranius and the passes. At first Caesar's army appeared to be heading towards Ilerda and by the time their true intentions were realised they were level with the camp of the Pompeians. Afranius abandoned his baggage and, leaving a guard of several cohorts in the camp, attempted to reach the mountains first, but he was seriously delayed by Caesar's cavalry constantly harrying his rear and the Caesareans won the race. Seeing Caesar's forces deployed for battle on a level stretch of ground before the passes, Afranius halted on a hill and sent four cohorts of targeteers towards the highest peak in sight with orders to occupy it. This advanced guard was attacked by Caesar's cavalry and surrounded. They held out for a considerable time but were eventually all killed.

Afranius then went into camp, followed by Caesar, who left cohorts to block all the passes. The Pompeians were now very short of food, had difficulty obtaining water and large numbers of the auxiliaries began to desert to Caesar. After several days Afranius broke camp to return to his supplies at Ilerda. However, he was again closely followed by Caesar who, whilst harassing him continually, did not seek a major confrontation. This harassment prevented the Pompeians from finding a suitable camp site and when they camped that night they had no water supply nearby. It was to be their last camp.

For another three days the two armies menaced each other from these camps and on the second day Afranius was forced to lead out his entire army in order to fetch water. By the fourth day his men needed both water and food, while the surviving animals had been without fodder for four days. Unable to reach either water or food, unable even to move camp, the Pompeian army surrendered. The leaders were spared and the entire army discharged.

The third of Pompey's generals in Spain, Varro, had two legions but after the defeat of Afranius the cities of Hispalis, Italica, Corduva and Gades closed their gates to him, one of his legions (a 'native' one) deserted, and in the middle of August Varro surrendered the other to Caesar.

CONCLUSIONS. For political reasons Caesar avoided a major battle in which thousands of Roman soldiers might have been slain by Roman hands. By intelligent construction of camps, bridges and a ford he not only cancelled out all Afranius' initial advantages but actually reversed the position and forced Afranius to abandon a position of great strength in order to maintain links with Varro and obtain adequate supplies.

FIRST SPANISH CAMPAIGN:
ILERDA

First Spanish Campaign: Ilerda

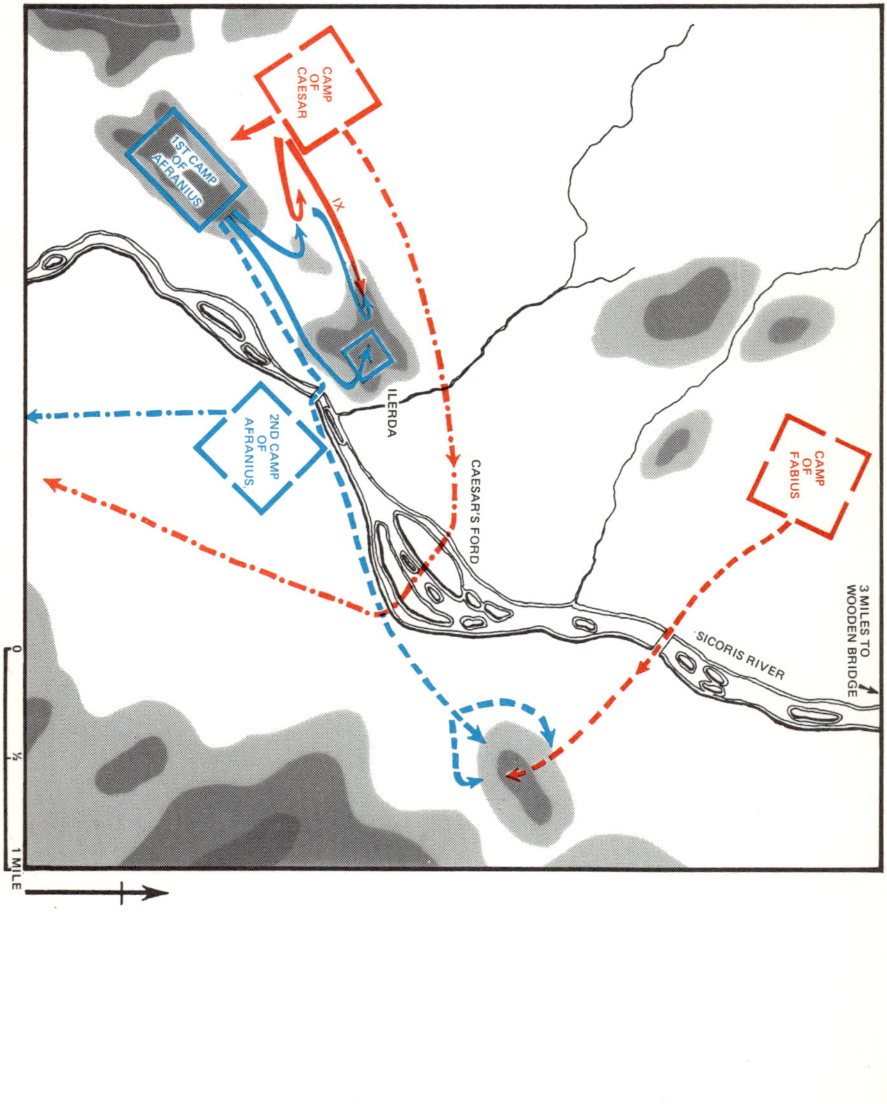

N

ILERDA The photograph shows the battle before Ilerda itself, one of several interesting games possible from this campaign. Caesar's cohorts have just been defeated in their attempt to gain the middle hill and may be seen retreating in the centre of the picture. IX Legion in the background is advancing to their aid. The spur on which the IXth became bogged down is on the extreme right, with Pompeian cohorts already moving in that direction.

FIRST AFRICAN CAMPAIGN

May – July, 49 B.C.

In the early spring Caesar had sent Gaius Curio to secure Sicily and this had been achieved without striking a blow by April 23rd. Two of the legions captured at the fall of Corfinium were then despatched to Sicily and Curio was ordered to take the African provinces for Caesar.

CAESAREAN FORCES. Commanded by Gaius Curio. Two legions of some 11,000 men with 500 cavalry. **In Support:** a fleet of 12 warships and a large fleet of transports.

POMPEIAN FORCES. Commanded by Attius Varus. Two legions of about 11,000 men with 600 Numidian cavalry and 400 Numidian infantry. **Reinforcements:** Varus was reinforced in greater strength later by King Juba of Numidia, although the actual numbers involved are not mentioned. During the 2nd Punic War Numidia had been capable of raising over 6,000 infantry and 8,000 cavalry so it is not unreasonable to assume that Juba could probably field an army of 18,000 men or even more, about 9,000 of whom might have been cavalry. The column sent to relieve Utica is unlikely to have exceeded 3,000 men, while the advanced guard at the Bagradas under Saburra may only have been 2,000 strong initially.

TERRAIN. The camp of Varus occupied a strong position between the walls of Utica and the town's amphitheatre.* Other buildings reduced the approach to a narrow and difficult one. Curio in turn had a camp of great natural strength at Castra Cornelia. These two camps were separated by only three miles but a stream, up which the sea flowed for some distance, made the ground marshy for some distance and it was necessary to make a detour of six miles to reach one camp from the other. A small but narrow valley, with sides so steep that it was impossible to scale them and fight at the same time, is also referred to but the author has been unable to positively identify it, the only valley which is identifiable being that referred to as marshy. This has made it impossible to show accurate troop movements on the map.

Curio landed at Cap Bon then sent his fleet to blockade Varus' camp at Utica while he marched the legions to the River Bagradas. Leaving the legions at that river, Curio took the cavalry and scouted as far as Castra Cornelia, where a camp had been constructed 200 years earlier during the 2nd Punic War. At the same time he observed the camp of Varus and noted the roads and tracks into Utica were crowded with refugees from the surrounding countryside. He turned his cavalry loose on these refugees and when Varus sent the Numidian cavalry and infantry to intervene the Caesarean cavalry charged them too and put them to flight. The Numidians lost about 120 men in this skirmish. Curio's warships now arrived and about 200 supply ships lying off Utica were informed they would be considered enemies if they did not at once transfer to Castra Cornelia. This they did with alacrity, thus supplying Curio with all the provisions for which he could wish.

Curio then returned to his legions and the next day led them to Utica and began building a camp near the town. Before the camp could be completed it was reported that a large force of Numidian cavalry and infantry was approaching and shortly afterwards the head of this relief column was sighted. Curio sent his cavalry to hold up the Numidians while he recalled the

*There was an 'amphitheatre' and a 'theatre' at Utica, the latter being outside the town walls. Caesar refers to the camp site as being between the town walls and the 'theatre' in front of the town. Since Curio was to the east, I have assumed Caesar referred to the 'theatre' outside the walls. Some translations refer to this as a 'theatre', others as an 'amphitheatre'.

legions from their work on the fortifications to form a line of battle but, before the legions had time to arm and deploy, the cavalry, charging the Numidians before they had time to break from their marching column, had completely routed the enemy. The Numidian cavalry escaped unharmed but a great many of the infantry were killed.

The next morning Varus marched his legions out of their camp and offered battle. Curio also formed a line of battle before his camp but neither force, separated as they were by the small, steep-sided valley, made any further move and eventually both returned to their camps without a blow being struck. Both armies returned to their battle positions the next morning and again each waited for the other to advance so that they might retain the advantage offered by the terrain. Suddenly the whole of Varus' cavalry on his left wing, together with a number of light troops, was seen advancing down into the valley. Curio sent his own cavalry with two cohorts to meet this attack. Varus' cavalry broke at the first charge and fled in disorder, abandoning the light troops who were then surrounded and cut down to the last man. Curio then gave the order for a general advance but, seeing the fate of the light troops, the remainder of Varus' line turned and fled to their camp before a missile could be thrown. They were closely pursued by the Caesarean legions.

In the struggle to get through the narrow approach to their camp more Pompeians perished than had been killed in the fighting and retreat combined, and many of the panic stricken troops continued their flight right through the camp and into the town itself. It was not possible to follow up this advantage because of the difficult approach to the camp and Curio therefore withdrew, having lost one man to Varus' 600 killed and 1,000 wounded. During the night the remainder of Varus' legions withdrew inside the town walls and citadel and in the morning Curio began constructing siege works round the town.

Shortly afterwards Curio learnt that King Juba was less than 30 miles from Utica, marching at the head of a second, much larger relief force, and he therefore withdrew to Castra Cornelia and began to strengthen the camp there, at the same time sending to Sicily for his other two legions and more cavalry. Later he received word that Juba had been recalled to deal with a threat to his own kingdom by the people of Leptis, leaving his general Saburra to march on Utica with a much reduced force. Late that evening Curio despatched his cavalry towards the Numidian camp by the Bagradas and, attacking in the middle of the night, these cavalry killed a great number of the sleeping Numidians and caused many others to flee.

Leaving five cohorts to guard his camp, Curio followed the cavalry at 0300 hours the next morning and after marching six miles encountered the returning triumphant cavalry. Their prisoners confirmed that it was Saburra and not Juba at the camp by the Bagradas and Curio therefore pushed on rapidly, hoping to catch the Numidians before they had time to reorganise. The cavalry was ordered to follow on behind but their horses were so exhausted by the all night march that most were unable to keep up.

Meanwhile Saburra had sent word of the cavalry attack to Juba, who was only six miles to the rear with the remainder of his army. Juba at once sent forward 2,000 Gallic and Spanish cavalry of his own bodyguard and the most reliable regiments from his infantry. He followed more slowly with the remainder of his forces, the baggage and 60 war elephants.

Saburra, suspecting that Curio's legions would be behind the cavalry attack, drew his forces into line of battle and gave them orders to pretend to be frightened and to give ground and retreat slowly until they were ordered to attack. Curio advanced about 11 miles in pursuit of Saburra because of this ruse, moving off the high ground on to the flat plain and exhausting his men unnecessarily. He then halted to rest his men but Saburra now gave the order to his forces to switch to the offensive and, keeping his infantry in reserve for the time being, launched a massive attack with his cavalry.

Only 200 cavalry had been able to keep up with Curio and although they were able to drive off the Numidian light cavalry whenever they attacked, they were too few to attack everywhere at once and their horses were too tired to pursue the enemy once they had started to retreat. On the other hand the numerous Numidian horsemen were able to outflank Curio's line of battle and began to attack the legionaries from the flanks and rear with a hail of javelins. On several occasions a cohort was detached from the line in an attempt to drive off the cavalry and these did succeed in their aim but when they sought to return to the main line they found themselves cut off and were slowly wiped out by the cavalry's arrows and javelins. The main body of the army, meanwhile, was forced into a tight circle where the legionaries had no room to swing their weapons and where the dead and dying remained upright because of the press. It was no longer possible for Curio to stand and fight, yet it was not safe to advance to the attack, nor to withdraw.

Juba was continuing to send forward reinforcements as fast as possible so that Saburra's strength was increasing every minute while Curio's men were becoming exhausted and demoralized by the constant attacks of the Numidian cavalry, the sheer weight of the weapons thrown against them being sufficient to force them slowly inwards on their own lines all the time.

Seeing that there could be only one end to this situation, Curio ordered the legions to make for the nearest hills but the Numidian cavalry was able to race ahead of the legions and forestall this move. This was the last straw for the legionaries and many attempted to flee while others dropped from exhaustion. These men were all killed by Saburra's cavalry. At last the tight mass of the legions broke up and the entire command scattered and ran, completely at the mercy of the Numidian cavalry. A few of the surviving Caesarean cavalry managed to escape to Castra Cornelia, together with those horsemen who had lagged behind the legions, but Curio and every man in his legions was slaughtered.

At Castra Cornelia there was panic when the survivors broke the news of the defeat. A few managed to escape with the fleet and transports but the vast majority of the five cohorts left at the camp and the surviving cavalry surrendered to Varus. Juba marched into Utica the next day and seeing these troops mustered in front of the town had most of them killed, keeping only a few as slaves. Varus lacked sufficient strength to prevent the massacre and could only voice his protests.

CONCLUSIONS. Caesar had given Curio four legions but Curio's opinion of Varus was so low that he considered two sufficient to defeat him. This was a mistake and Curio also failed to make any allowance for the intervention of Juba, who might have been expected to side with Pompey since Pompey had in 81 B.C. restored Juba's father to the throne while Curio as a tribune of Caesar had only a year before reduced Numidia to a form of Roman province. After a good start to his campaign Curio made his second and fatal mistake by rushing at Saburra without carrying out sufficient reconnaissance to discover what forces were before him, at a time when his own cavalry were exhausted. Had he waited in the safety of Castra Cornelia for his other legions he could have worn Juba's strength down before making a major attack under more favourable circumstances. As it was his impetuousity cost over 11,000 men their lives and the loss of Africa gave the defeated Pompeians a refuge from Caesar's legions in the next year.

CASUALTIES. (for the campaign) **Caesarean:** Curio's entire army of 11,000 legionaries and 500 cavalry was lost. **Pompeian:** 600 legionaries killed, 1,000 wounded. About 300 Numidian cavalry and 1,400 infantry were killed.
(Numidian figures are the author's, based on various accounts.)

FIRST AFRICAN CAMPAIGN

First African Campaign

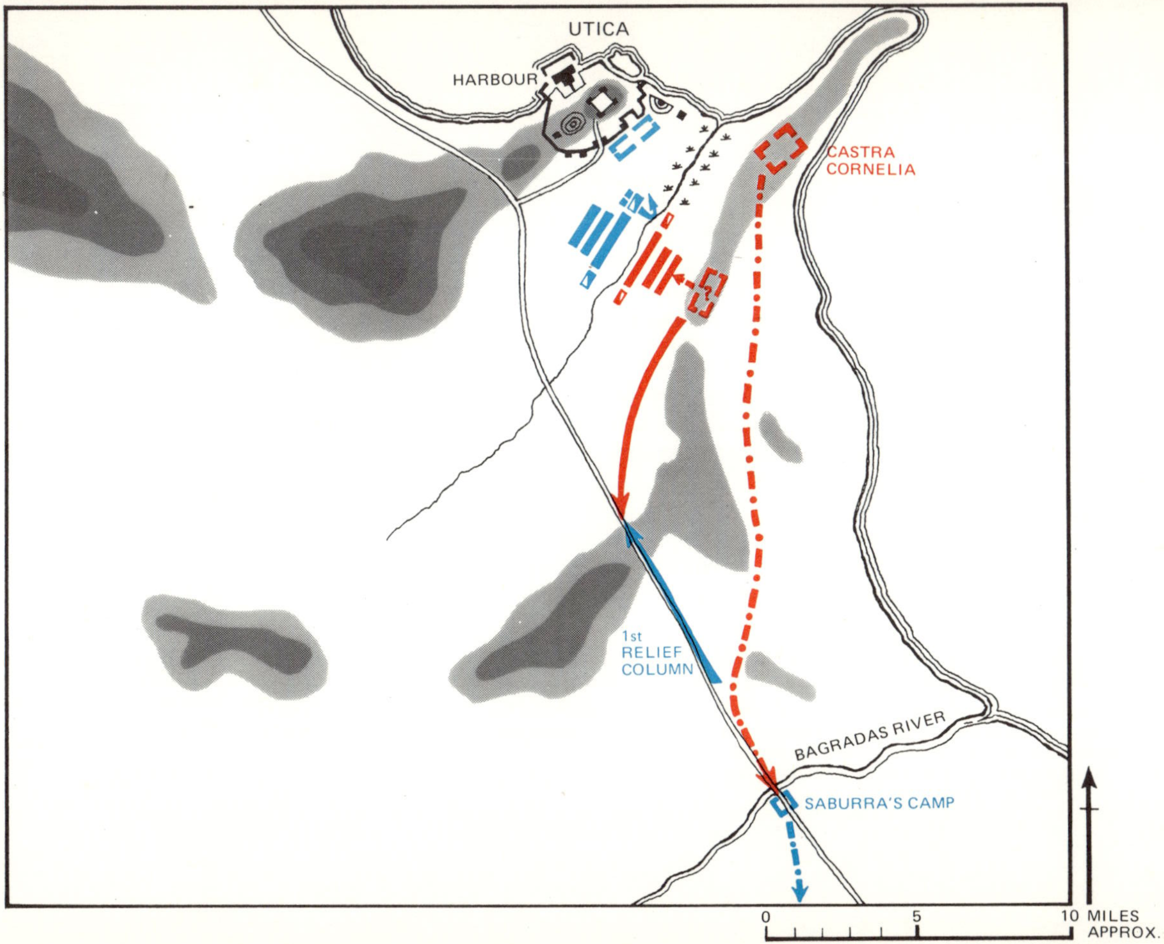

N

1st. AFRICAN CAMPAIGN The defeat of Curio by Saburra and Juba, showing the massive Numidian light cavalry attacks against the solid yet helpless ranks of the tired Roman legions. Saburra's infantry are well to the rear and more cavalry are continually arriving from Juba, further back and off the board.

DYRRACHIUM

May 25th, 48 B.C.

Caesar did not have time to retrieve the position in Africa. Returning from Spain he put down a mutiny at Placentia and, learning he had been named dictator, went to Rome to hold consular elections and pass several important laws. Although his officers had been gathering and building ships all the time he had been in Spain, Caesar still did not have sufficient vessels to transport the whole of his army to Greece. On January 4th 48 B.C., therefore, he sailed from Brindisi with only half his army and landed at Oricum, where the town surrendered to him. There then developed a race between Caesar and Pompey to reach the latter's magazines at Dyrrachium. Caesar won and built a camp which cut Pompey's army off from its arsenal of weapons. Pompey then built a strong camp at Petra, which had a good natural harbour, and ordered corn to be conveyed there from all the regions he controlled.

Caesar seized a number of hills around Petra, built forts on them and began linking the forts with earthworks, attempting to enclose Pompey in a small area around Petra and starve him into submission. Although Pompey's army was superior in numbers it was far from being the equal of Caesar's veterans and Pompey therefore refused to be drawn into a pitched battle to prevent this work. At the same time he did not wish to leave the vicinity of Dyrrachium, nor abandon his position by the sea, over which his navy had control. This left only one course, to form a line of entrenchments within the line of contravallation being built by Caesar. This led to a second race, each army trying to seize as much territory as possible, Pompey in order to obtain a larger area for foraging, Caesar to deny him this. Having a larger army and a smaller circuit to build, the Pompeians won this race, while the outer line of fortifications stretched Caesar's manpower to its limit. (Caesar's lines were 15 miles long and were the most extensive ever constructed in classical times.) Nevertheless, Pompey's army, particularly the cavalry, soon began to feel the effects of the encirclement and towards the end of May Pompey began to seek a means of breaking out. At this point two cavalry officers of some importance deserted to Pompey and pointed out a weak part of Caesar's line – the unfinished entrenchments at the southern extremity by the sea. Pompey took the offensive.

CAESAREAN FORCES. An initial garrison of two cohorts from the IX Legion camped by the sea shore, reinforced soon after the fighting began by several cohorts and some cavalry from the Legion's camp. (About half the IXth was committed before retreating into their camp.) **Reinforcements:** 12 cohorts under Mark Antony, arriving as Marcellinus' camp was threatened. A total of 35 cohorts, including the weakened IX Legion and Antony's men, and about 700 cavalry were involved in the second phase of the battle led by Caesar himself.

POMPEIAN FORCES. Six legions led by Pompey and assisted by sea-borne landings by a 'great number' of light troops and archers. The author has only been able to identify about 3,000 archers and 1,200 slingers to represent this 'great number'. **In Support:** up to 7,000 cavalry.

TERRAIN. Caesar had been worried that this southern extremity of his line might be attacked and outflanked by a sea-borne force and had therefore built a double line of fortifications here. Facing north was a rampart 10 feet high and 10 feet wide across the top, fronted by a 15 foot wide ditch. 200 yards to the rear of this was a second, lower rampart, facing south. A cross rampart, facing the sea, was being built but had not been completed. In addition to Marcellinus' camp there was a second fortified camp in the area. This had initially been built by Caesar between the lines of the two armies but had later been abandoned. Pompey had extended this fortification to house several legions, incorporating Caesar's camp as an inner citadel, but had also abandoned the position. Thus, at the start of the battle, neither side had possession of this fortification.

Pompey's light troops were embarked the evening before the battle in dinghies and fast boats with orders to land in two parties on the flank and rear of Caesar's lines. They had made wicker coverings for their helmets and took with them brushwood to fill the enemy's ditch and stakes to build a new parapet on the other edge of the captured rampart. Just after midnight Pompey led six legions, drawn from his main camp at Petra and surrounding outposts, to that part of his line nearest the unfinished cross rampart and furthest from Caesar's main camp between Petra and Dyrrachium.

The legions attacked the north face of Caesar's double line at dawn. Simultaneously most of the light troops were landed in the rear of the line and began throwing missiles at the defenders on the southern rampart and filling in the ditch with brushwood and earth. Whilst archers spread round the end of the double line and poured in arrows from every direction the legions brought forward scaling ladders and ballistae. The defenders had only stones with which to reply to this hail of missiles, and these had little effect because of the wicker covering on the Pompeians' helmets. As they were struggling to hold on to their lines of defence another body of light troops disembarked at the point between the two ramparts, splashed through the water and attacked both lines of defenders from behind.

The survivors of the two cohorts of the IXth were driven from the walls and began to fall back towards the legion's camp. Marcellinus had by now received information of the attack and had despatched some cohorts (perhaps two or three) and some cavalry to help hold the walls but these troops were unable to halt the enemy's advance. Further reinforcements were sent forward but by now the IXth was in flight, the reinforcements were infected by the general panic, and the Caesareans fled towards their camp, losing many men in the process.

Pompey now advanced on Marcellinus' camp, where the remaining cohorts of the IXth waited in fear, and was about to storm the camp when Mark Antony, commanding the next nearest post, was seen marching down from the high ground to the north east with 12 cohorts. Antony's force drove the Pompeians back a little and Pompey therefore gave orders to construct a fortified camp near the coast. This camp would ensure he was able to keep Caesar's lines broken and so allow the starving cavalry horses to graze in fresh areas.

The IXth now rallied and before long Caesar, who had been informed of the situation by smoke signals sent from one fort to another, arrived on the scene with about 15 cohorts. Seeing that his lines had been pierced, Caesar ordered a camp to be built near Pompey's and constructed a cross rampart between his lines near that point. As the camp was being completed scouts reported that a legion was using the cover of the wood to occupy the abandoned camp about half a mile to the north. Hoping to recuperate his losses, Caesar decided to strike at this legion before Pompey could support it and, leaving two cohorts behind to guard the new works and give the impression that work was continuing on the fortifications, he led the remaining 33 cohorts in two lines abreast against the northern camp.

The Caesareans reached the camp before Pompey was aware of the manoeuvre. (The maps do not show any contours which could have concealed this move. It is possible the ramparts themselves, both of the two new camps and the original lines of circumvallation – the rampart of which you will remember was 10 feet high, concealed this manoeuvre for most of the distance covered.) Attacking quickly with his left wing Caesar forced the Pompeians from the ramparts but found the gates barred by a 'hedgehog', a wooden beam studded with sharp spikes. After a short tussle the hedgehog was hacked apart and the Caesareans burst into the camp and, following up their success swiftly, also pierced the defences of the inner camp, killing a good number of defenders.

Unfortunately Caesar's right wing, not knowing the area, had been side-tracked by the wall extending towards the river, built to provide a covered way to water. Believing this wall to be part of the camp they followed it northwards, looking for the gate. Discovering that it reached the river without having a gate, and that the wall was undefended, they broke through the wall near the river and passed inside Pompey's lines, followed by all the cavalry.

Pompey had at last received news of the attack. Withdrawing five legions from his new camp he marched rearwards to assist the isolated legion. At the same time his cavalry was seen approaching from the north. The defenders of the northern camp were so encouraged by the sight of all these forces hurrying to their aid that they rallied by the western gate of the smaller fort and made a counter-attack. The Caesarean cavalry, making its way along a narrow path over the earthworks and afraid of being caught in such a cramped space, now turned and fled towards the rear. The cohorts of the right wing, cut off from the left and seeing their cavalry in retreat, also began to withdraw hastily, afraid of being trapped within the confines of the fortifications by attacks from two sides. In the resulting panic and confusion many were killed by falling from the ten foot high rampart into the deep ditch, where they were trampled by those behind. The left wing, observing the rest of the army in flight and afraid of being attacked from inside the camp by its defenders and from the outside by Pompey's legions, also began to pull out. The advance of Caesar's victorious forces had suddenly disintegrated into confusion, panic and rout.

Caesar attempted to rally his cavalry outside the ramparts of the camp to cover the withdrawal of the legions, actually taking hold of some of the standards in an attempt to halt their bearers. Some bearers dropped the standards when thus accosted and continued their flight, one even struck at Caesar with his sword, a blow parried by Caesar's shield bearer. But luck had not completely deserted Caesar. Unable to believe that such a sudden reversal of fortune did not smack of a trap, Pompey ordered a halt. The Pompeian cavalry was hindered by the entrenchments and narrow gates and so was unable to conduct a proper and effective pursuit. These two circumstances enabled the routed Caesareans to escape to their camp with relatively slight loss.

CONCLUSIONS. Caesar was able to rush reinforcements to the weak spot very quickly but Pompey, by building a camp south of Caesar's lines, had already rendered the whole of Caesar's double line of fortifications in the south obsolete. Caesar's attempt to salvage something by attacking the legion in the northern camp was fated once his right wing was side-tracked by the river wall and became bogged down in Pompey's lines. These same fortifications did, however, save the Caesareans from annihilation by preventing the Pompeian cavalry from deploying in time to pursue them. Pompey's five legions might have been able to pass through the confusion in the camp and decimate the routed enemy but, fearing an ambush, he stopped his advance at the moment of victory, and subsequently withdrew into his new camp. Despite Pompey's failure to exploit his success, Caesar's grip on the Pompeians was shattered and shortly afterwards he abandoned his lines and headed for Apollonia, planning to march into Macedonia against the forces under Scipio. He hoped by this manoeuvre to give his shaken army time to recover and at the same time be in a position to defeat Scipio before he could be supported by Pompey, or at least to draw Pompey away from the sea and bring him to a major confrontation on equal ground.

CASUALTIES. Caesarean: 960 infantry killed, 200 cavalry killed. Many senior officers, 32 centurions and tribunes were lost, together with 32 standards. Most of the casualties were caused by the panic stricken flight, not by enemy weapons. Many prisoners were also slaughtered after the battle by the Pompeians. **Pompeian:** unknown, but since only the legion in the northern camp was heavily engaged the numbers are unlikely to have been very large, perhaps as few as one or two hundred.

DYRRACHIUM

Dyrrachium

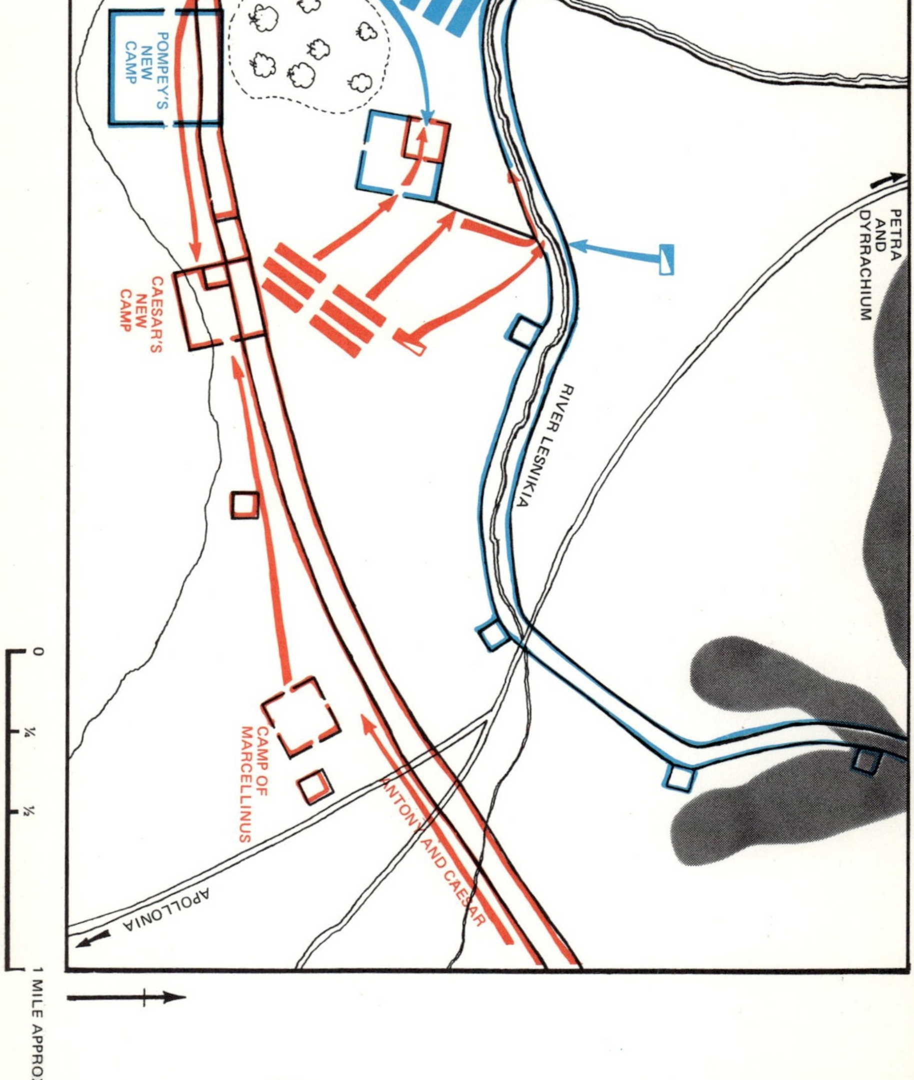
POMPEY'S NEW CAMP
CAESAR'S NEW CAMP
CAMP OF MARCELLINUS
ANTONY AND CAESAR
APOLLONIA
RIVER LESNIKIA
PETRA AND DYRRACHIUM
0
¼
½
1 MILE APPROX.

N

DYRRACHIUM This photograph shows Caesar's attack on the old fort between the two siege lines, the action being fought as a separate wargame with Pompey's five legions arriving halfway through the game. The right wing and cavalry are just moving into their fatal error which cost Caesar the battle in real life and, having made this move, the game was taken from there. The ramparts of Pompey's siege lines in the background were made from Plasticine: the fort is the Bellona Roman Fort.

PHARSALUS

June 29th, 48 B.C.

After his defeat at Dyrrachium Caesar marched into Thessaly where he had stationed the XI and XII Legions and 500 cavalry under Domitius Calvinus to intercept Metellus Scipio, governor of Syria, who was bringing two legions and 200 mounted Syrian archers to join Pompey. Domitius was almost caught between Scipio and Pompey but was warned just in time by Caesar and managed to escape to unite his force with Caesar, but in the process Caesar was unable to prevent Scipio and Pompey from joining forces. Caesar was anxious to win a decisive victory but Pompey had been content to avoid a major battle and attempt to weaken Caesar's army by cutting his lines of communication and depriving him of supplies. The two armies now arrived on the plain of Pharsalia where Pompey's commanders demanded a council of war at which it was decided that a decisive battle should be fought.

CAESAREAN FORCES. 80 much understrength cohorts, including the VIII, IX, X, XI and XII Legions; a total of about 22,000 legionaries with 1,000 cavalry and a small number of light troops.

POMPEIAN FORCES. 110 cohorts of legionaries and auxiliaries totalling 45,000 men and made up from the I and III Legions (under Caesar they had been the VI and XV); the Cilician or Twin Legion which was made up of veterans from two legions which had served under Cicero in 51 – 50 B.C. and reinforced by some Spanish cohorts brought over by Afranius; 2 Syrian legions; 5 Italian legions; 2 Asian legions; and a legion of veterans from Crete and Macedonia. Amongst the auxiliary cohorts were men from Numidia, Thessaly, Gaul, Spain, Boeotia, Achaea and Epirus. The legions were supported by 3,000 archers from Crete, Sparta, Pontus and Syria, 1,200 slingers, and 2,000 veterans of a special corps **(beneficiarii).** The cavalry comprised 600 from Gaul, 500 from Cappadocia, 500 from Thrace, 200 from Macedonia, 500 Gauls and Germans, 800 of Pompey's herdsmen and slaves, 300 from Gallograecia, 200 Syrian mounted archers, 600 from Pontus and the remainder from Dardani, Bessi and Thessaly to a grand total of 7,000 men.

TERRAIN. The main fighting took place on the flat plain of Pharsalia but the end of the battle came in the hills which surrounded the plain on three sides.

When the two opposing armies camped on the plain of Pharsalia they were separated by a distance of four miles. For some days the two generals marched their armies out of camp and formed line of battle but Pompey always drew up his line on the lower slopes of the hill on which his camp was situated and Caesar therefore did not take the offensive. Eventually, beginning to run short of corn, Caesar decided to move on, hoping to exhaust Pompey's army with long marches, but even as the head of his column reached the camp gates on June 29th it was seen that Pompey had that very morning drawn up his line of battle much further from his camp than usual, offering battle at last in a position which gave advantage to neither side.

Caesar rode towards Pompey's line to study the enemy's dispositions while his own legions formed their customary three lines. Pompey was using the River Enipeus (Tchinarli), which had steep banks, to protect his right flank and had concentrated all his cavalry, slingers and archers on his left, with the exception of about 600 Pontic and Cappadocian cavalry which were on his right. At the extreme left of his line were the I and III Legions, first class fighting units under Pompey's own command. His extreme right was held by the Cilician Legion with the Spanish cohorts. The Syrian legions were in the centre and the remainder of the cohorts were distributed between the two wings and centre with some of the auxiliaries interspersed between the lines to act as light infantry. Seven cohorts had been left to guard the camp and several smaller forts on nearby hills. This gave Pompey about 42,000 infantry against Caesar's 21,500 (Caesar had left only two cohorts to guard his camp) and 7,000 cavalry to Caesar's 1,000. This greater number of cohorts meant that Pompey's frontage was longer than Caesar's.

Caesar united the weakened IX with the VIII Legion and placed them on his left, and personally took command of the X Legion on the extreme right. Seeing that his right was certain to be outflanked by the enemy's cavalry and light troops, he withdrew six cohorts (according to Appian it was in fact 3,000 men) from the reserves of his third line and placed them in a fourth line on his right behind the Xth with orders to keep out of sight until ordered to charge. The reserves of the third line were also ordered not to become involved in the fighting until they received the order to charge forward from Caesar himself.

There was sufficient space between the two armies for both to advance to the attack at the double but Pompey ordered his men to stand firm and receive the enemy's charge at the halt. He hoped by this tactic to have the Caesareans arrive out of breath and with their line broken as it spread out to encompass his longer one, thus breaking up any charge impact. It was also hoped the enemy's javelins would be less effective against a shield wall, while their own would be more effective. But Pompey had not allowed for the battle experience of Caesar's legions. As soon as the signal to attack was given the whole line surged forward eagerly but, noticing the enemy had not moved, the legions checked their advance at about the half way mark without orders and paused for a while to regain their formation and breath. They then charged forward again, hurled their javelins and, drawing their swords, moved in.

The Pompeians received the javelins on their shields, threw their own in reply and received the charge without breaking ranks. At this moment the mass of cavalry on Pompey's left wing charged forward, closely followed by over 4,000 archers and slingers. The Caesarean cavalry, strengthened by light troops intermingled amongst its ranks, did stand up to this onslaught but stood no chance at all against such overwhelming numbers and was soon forced to give ground. Immediately the Pompeian light troops began to assail Caesar's right with missiles while their cavalry deployed by squadrons to sweep round the exposed flank. At this moment Caesar gave the signal to his fourth line of six cohorts, and these charged forward to attack the cavalry. Instead of throwing their javelins these cohorts used them to thrust upwards at the faces of the cavalrymen. Taken by surprise, fighting their first battle, composed of young handsome nobles proud of their good looks, the front ranks of the cavalry sheered away, unable to face the hedge of steel points flickering towards their eyes and mouths. Abruptly the whole mass turned away and not only gave ground but became infected by panic and fled from the field of battle to seek refuge in the hills. Caesar's fourth line paid them no more attention but fell instead upon the unarmoured archers and slingers who, unsupported and defenceless, were slaughtered to a man.

In the main line of battle Pompey's legions and auxiliaries were still standing firm and no advantage had been gained by either side, but now Caesar's fourth line swept round the flank of the main line and began to attack the left flank from the rear. At this point Caesar ordered his third line forward to replace the exhausted troops of the first and second lines. Attacked by fresh veterans in front and by others on the flank and to the rear, Pompey's left wing began to break up. The centre fought on as stoutly as ever but Pompey himself now quit the field and rode to his camp, where he made a pretence of alerting the camp guards. He was followed soon after, at about mid-day, by disorganised and defeated cohorts of his army, which retreated into the camp, many of the men being so exhausted and demoralized that they abandoned their weapons and even their standards. The cohorts guarding the camp put up a stout resistance, as did the Thracian and other auxiliaries with them, but Caesar pressed home his success fully and, heavily outnumbered, the camp defenders were overwhelmed by a hail of missiles and abandoning the ramparts retreated towards the safety of the high ground above the camp with most of the army.

After occupying Pompey's camp Caesar ordered his tired men forward again and began constructing earthworks round the hill to which the bulk of the enemy had fled. The hill was without a water supply and seeing the position was hopeless the Pompeians began to retreat again, moving eastwards along the ridge of the hills. Caesar ordered some of his cohorts to return to his camp, others to Pompey's camp, then with the remaining four legions crossed the Enipeus and marched between the hills to cut off the enemy's retreat. Because the Pompeians were making their way over the hills, Caesar was soon able to draw ahead of them and after a march of four miles he formed a line of battle across their line of retreat. The Pompeians, seeing these legions waiting for them, halted on the high ground, on a hill along the foot of which ran the Enipeus. It was now almost dark and both sides were utterly exhausted but somehow Caesar managed to get his legionaries to build a line of fortifications between the high ground and the river, so cutting the Pompeians off from any water. At dawn the next day the entire surviving forces of Pompey surrendered to Caesar.

CONCLUSIONS. Pharsalus was the most bloody battle of the Civil Wars as far as Roman versus Roman was concerned and the thousands of Romans killed in this battle affected the development of the nation for many decades. The battle cost Pompey not only control of Greece but also his life, for instead of heading for Tunisia, where Varus and his ally Juba had been victorious, Pompey took ship to Egypt, believing he might raise a new army there. He was assassinated as he stepped ashore. Caesar demanded a great deal from his legions that day, and got it. By pressing the pursuit of the defeated Pompeians to the bitter end he obtained his decisive victory, and with Pompey dead there was no longer anyone to seriously oppose him. When news of Pharsalus reached Rome he was declared dictator for an indefinite period and most people believed the Civil Wars had ended.

CASUALTIES. Caesarean: 200 men and 30 centurions killed according to Caesar. Other sources state 1200. **Pompeian:** 15,000 killed on the field or during the retreat, probably as many as 6,000 legionaries in the fighting in and around Pompey's camp alone. 24,000 surrendered, including the garrisons of the detached forts, and the remainder of the army was so scattered as to be useless for the future. 180 standards and 9 eagles were captured and Domitius Ahenobarbus was killed whilst attempting to escape from the camp to the higher ground.

PHARSALUS

Pharsalus

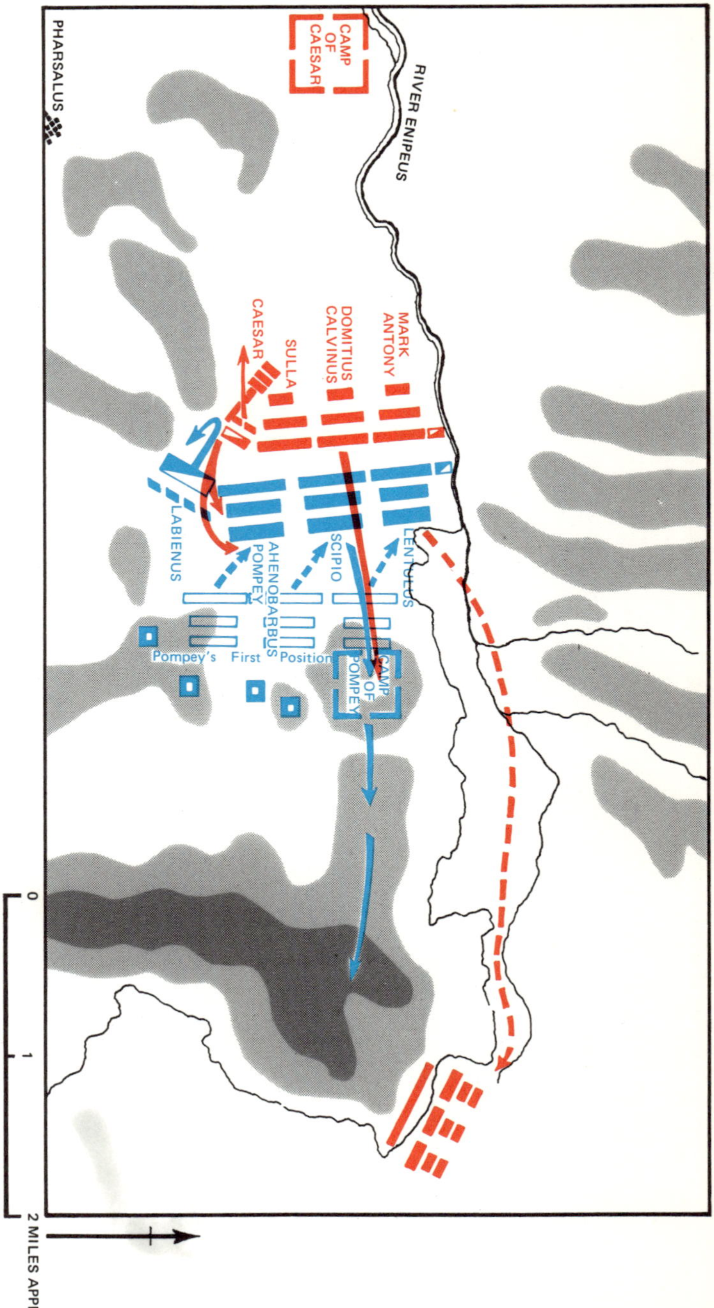
PHARSALUS
CAMP OF CAESAR
RIVER ENIPEUS
MARK ANTONY
DOMITIUS CALVINUS
SULLA
CAESAR
LABIENUS
POMPEY
AHENOBARBUS
SCIPIO
LENTULUS
Pompey's First Position
CAMP OF POMPEY
0
1
2 MILES APPROX.

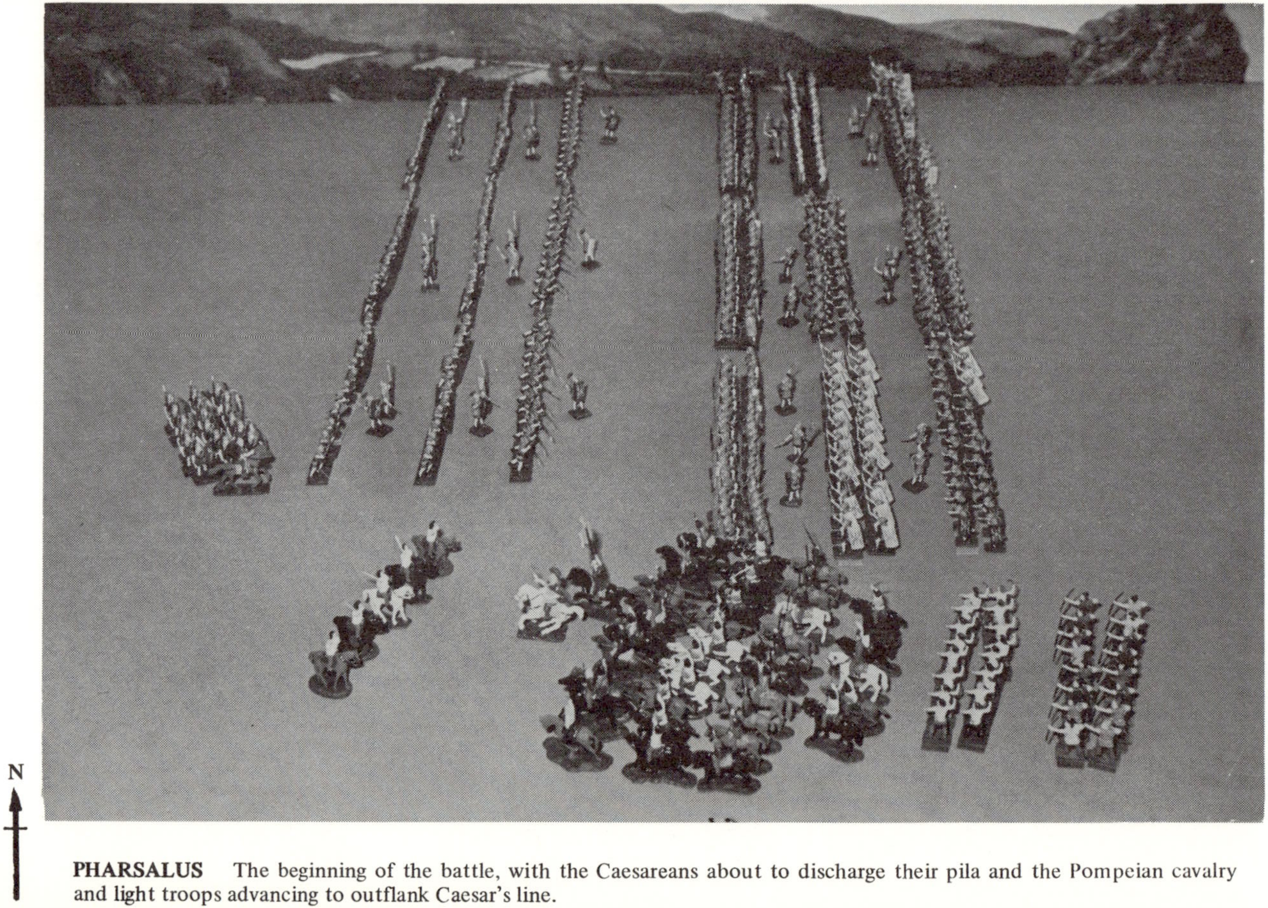

PHARSALUS The beginning of the battle, with the Caesareans about to discharge their pila and the Pompeian cavalry and light troops advancing to outflank Caesar's line.

SECOND AFRICAN CAMPAIGN: RUSPINA

January 4th, 46 B.C.

Due to intrigue in Egypt, which considerably delayed Caesar, Scipio and Cato escaped to Africa after Pharsalus and there joined with Attius Varus and King Juba to assemble a formidable army. When Caesar finally returned to Rome late in 47 B.C. it was to find that a second campaign was vital in Africa. He set out for Africa in December but his fleet was dispersed by a storm and he arrived at Hadrumetum with such a small force that he was forced to march southwards and on January 1st camped at Ruspina. The next day he proceeded to Leptis Minor, which agreed to side with him, and he made camp by the shore not far from the town. Here he was joined by some of his transports and warships. Leaving six cohorts to garrison Leptis he returned to Ruspina on the 3rd and entrenched the remainder of his forces here. That night his missing troops arrived at that very town and therefore on the 4th Caesar set off with 30 cohorts to gather corn from the surrounding countryside.

CAESAREAN FORCES. 30 cohorts, including some from the X Legion, a total of about 14,500 men, many of whom were new recruits, with 150 archers and 400 cavalry. Total, 15,050 men.

POMPEIAN FORCES. Commanded by Labienus. About 35,000 heavy and light infantry, 7,000 Numidian cavalry (without bridles), 1,600 Gallic and German cavalry. **Reinforcements:** 1,600 elite Numidian cavalry and perhaps 2,000 Numidian infantry under Petreius. Total, 47,200 men.

TERRAIN. The action was fought on a small plain some three miles from Ruspina. Caesar himself refers to the plain as being absolutely flat but mention is made elsewhere of high ground to the front and flanks nearby, though only one small hill features in the actual fighting.

After marching for about three miles from Ruspina scouts informed Caesar that enemy forces were approaching and soon after this a large cloud of dust became visible. Caesar advanced to reconnoitre with a few cavalry, ordering the remainder of his force to follow slowly, and after a while was able to see the forces of Labienus drawn up in an extremely long yet closely packed line. Although it must have been obvious that he was outnumbered by three to one, Caesar formed his men in a single line with the archers in front and the cavalry on the wings and offered battle.

The two battle lines then closed on each other and halted when they were a short distance apart. Caesar could now see that the enemy's centre was not composed of heavy infantry as might have been expected and as he had assumed, but of Numidian cavalry interspersed with Numidian light infantry and unmounted archers. In view of this unusual formation, the vast numbers of Numidian cavalry and the general numerical superiority of the enemy, Caesar stood fast and made no attempt to attack.

After a while the cavalry on the flanks of Labienus' line began to spread out to the open flanks and take in the surrounding high ground. This forced Caesar's cavalry to spread out very thinly in order to prevent encirclement and in the light skirmishing which then took place the Caesarean cavalry was hard put to hold their own.

It was now 1100 hours and the two centres had drawn near enough to each other to become engaged, when suddenly the Numidian cavalry and light infantry charged forward in a mass and discharged a hail of javelins and arrows at Caesar's cohorts. The legionaries responded by charging home but the Numidian cavalry wheeled swiftly away, the light infantry stood firm for a while, and in attempting to pursue the cavalry and deal with the light infantry the

Caesarean line became broken, allowing the light troops to thrust javelins into the exposed sides of the legionaries. The cavalry then returned to catch the cohorts out of formation, darting away again whenever the legionaries tried to attack them and leaving them to be attacked in turn by the light infantry. Seeing the futility and danger of this mode of fighting, Caesar passed the order along the ranks that no man was to advance more than four feet in front of the standards.

By now the Caesarean cavalry was becoming exhausted in attempting to drive back all the thrusts made by the Numidian cavalry and also owing to great inferiority in numbers was being gradually forced back. This allowed the great mass of Numidian light cavalry to encircle Caesar's smaller force and with the light infantry they began making javelin attacks from all sides. Caesar's cohorts formed a circle and continued to fight stubbornly but; hemmed in on all sides and compelled to fight in a cramped space against hopeless odds, they rapidly began to lose heart and eventually did nothing except take shelter behind their shields and dodge the enemy's javelins as best they could.

What followed is not clear. Caesar says he ordered a double line formed, every other cohort facing about, and was thus able to extend his line to such a degree that the flanks broke through the mass encircling them. Since he was already in a circular formation it seems more likely that he changed this to an oval, then pushed outwards from the two ends to form a line facing in both directions. He then used his cavalry, presumably still close by and under control despite being outnumbered four to one, to help keep the two halves of the enemy forces apart while the cohorts launched an attack in both directions at once. The determination of the attack and its unexpectedness caused the Numidians heavy casualties before they managed to retreat out of range. Caesar then attempted to fall back on his camp before the Numidians could rally. (Labienus' heavy infantry do not appear to have been engaged at all and presumably only the Numidian light infantry and cavalry without bridles were involved in this close fighting.)

At this decisive point in the battle Marcus Petreius arrived on the scene with 1,600 elite Numidian cavalry and about 2,000 Numidian infantry. This force immediately attacked the withdrawing Caesareans, forcing them to abandon their withdrawal and form line of battle again, while the recently routed Numidian cavalry and infantry now rallied and returned to the attack. It was getting late in the afternoon, Caesar's cavalry were utterly exhausted and his infantry largely ineffective against the Numidian tactics. Caesar therefore went round his men, encouraging them to make one last effort to drive the enemy back beyond the high ground before them and to take possession of this height themselves.

The Caesareans waited until the Numidians became over-confident again then launched a sudden all out attack which took the enemy so much by surprise that they were driven right off the plain and the Caesareans were able to occupy the hill at the edge of the plain without any trouble.

The Caesareans had by now suffered many casualties and it was decided to make a last stand on the high ground in the hope that nightfall might enable them to slip away. This was a faint hope and they would probably have been cut to pieces by the Numidians had it not been for the fact that both Labienus and Petreius had been wounded, the latter seriously, and, lacking a Roman leader who could concert their attacks in the best method for engaging legions, the Numidians frittered away their chance of a decisive and important victory without making any combined effort to overrun the outnumbered and stranded Caesareans.

At sunset Caesar made a break for his camp but his cohorts were completely routed by the highly mobile Numidians. Fortunately for Caesar, Labienus' horse was wounded and threw him during this rout and the pursuit was then abandoned by the Pompeian forces.

CONCLUSIONS. Caesar does not refer to Ruspina as a defeat, merely stating that after holding the hill for a short while they withdrew in formation to their camp. This clashes directly with the other contemporary sources which state that the Caesareans were completely routed and many were killed.

Scipio was approaching with another eight legions and 3,000 cavalry to join Labienus and Petreius and Caesar now spent all his time strengthening the fortifications at Ruspina and making good his losses by arming Gallic and Rhodian rowers, Syrian archers, marines and many other men of different nationalities from the fleet. He then awaited the arrival of the remainder of his army and subsequently broke out of Ruspina, attempted to take the town of Veita, then marched on Thapsus and laid siege to it, forcing Scipio to attack him there. As a result of the fighting at Thapsus, and then Utica, Caesar once more gained control of Africa.

CASUALTIES. Not known for either side, but obviously heavy losses were suffered by the Caesareans in contrast to light casualties by the Numidians.

SECOND AFRICAN CAMPAIGN: RUSPINA

Second African Campaign: Ruspina

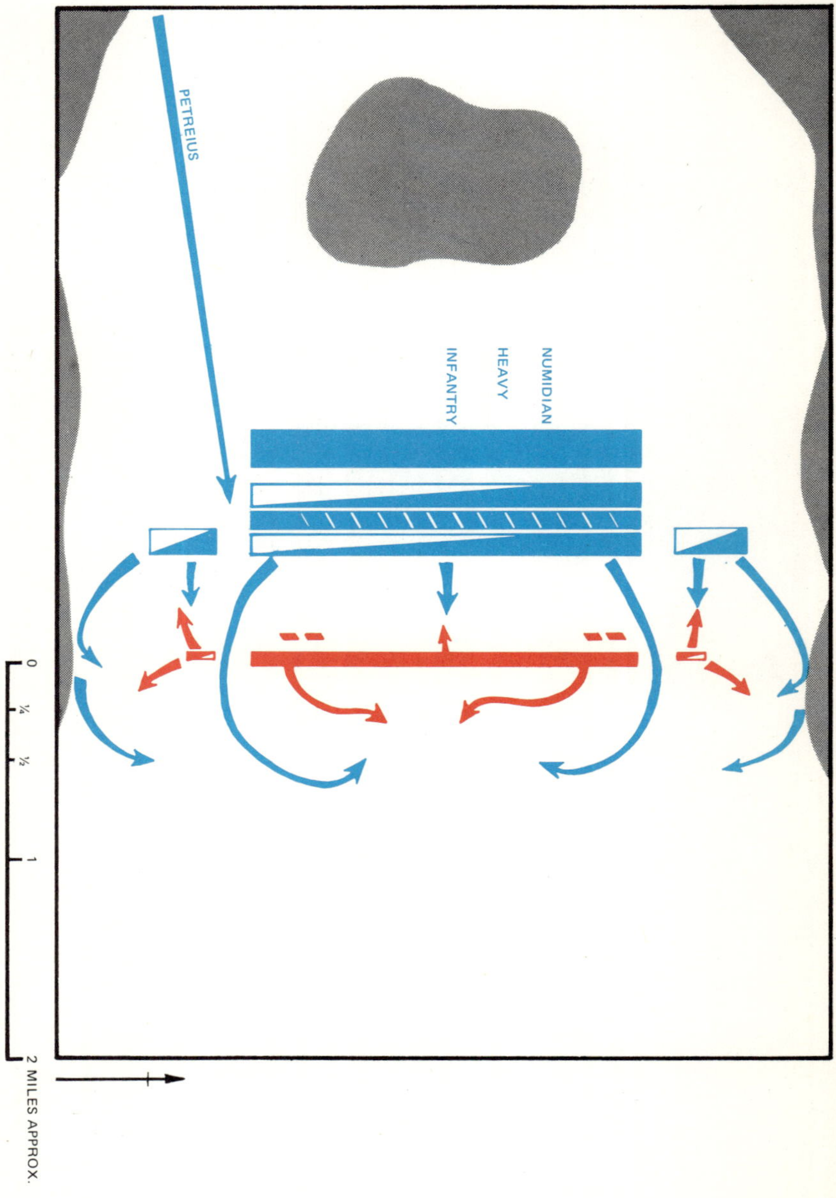

N ←

RUSPINA The scene at about 1100 hours, the beginning of the unequal engagement between Numidian light troops and cavalry and Caesar's heavy infantry. The Numidian heavy infantry is not on the board, the cavalry and light infantry in the foreground being fresh waves waiting to attack the Roman line with showers of javelins and arrows.

SECOND SPANISH CAMPAIGN: MUNDA

March 17th, 45 B.C.

After crushing the Pompeians in Africa Caesar returned to Rome in July, 46 B.C. However, after only five months he was forced to go to Spain to fight a second campaign there, this time against the sons of Pompey who, because of misgovernment by two successive governors appointed by Caesar, had been successful in starting an insurrection and were now at the head of a large army in Further Spain. After various minor engagements the Pompeians finally camped on the edge of a plain by the town of Munda, closely followed by Caesar, and it was here that the decisive battle of the campaign was fought.

CAESAREAN FORCES. 80 cohorts, including the III, V, VI and X Legions, totalling 32,000 men, with 8,000 Italian and Numidian cavalry and an unknown number of auxiliaries. Total, 40,000 men, plus the auxiliaries.

POMPEIAN FORCES. Commanded by Gnaeus Pompey. 13 legions, including the I, II, VIII, a native legion, one raised from Roman colonists of Further Spain and one of Afranius' old legions from Africa, the remainder being formed of auxiliaries and runaway slaves. Total, 44,000 men. There were also 6,000 light infantry and about 6,000 cavalry. Grand total, about 56,000 men.

TERRAIN. The site of the battle of Munda has not been precisely located but is generally believed to have been about 90 miles north east of Gibraltar. We know only that Munda was a fortified town on a hill in a high-lying region and that before it lay a plain some five miles wide. Caesar's camp was at the far edge of the plain from Munda. Because the site has not been identified no authentic map exists of the battle area. However, from contemporary sources it is possible to compile a rough map to serve as a guide to the battle itself. A better idea of the conditions in which the battle was fought might be obtained from the following precis of Caesar's own comments on the surrounding area:- The district was fertile, well watered and very hilly, with an occasional plain intervening. The terrain was therefore ideal for the construction of fortified camps, which could rarely be conquered by siege. Most of the towns in the province were protected against raids by the natives by being built on positions of natural strength, a steep climb to the town's gates making access difficult. Things were made even more difficult at Munda by a stream, one bank of which was marshy and full of bog pits, which ran along the foot of the hill on which the town was built.

Gnaeus Pompey camped his army under the walls of Munda on March 15th. Caesar arrived the next day and on the following morning was about to advance to besiege the town and Pompey's camp when the enemy were seen leaving their camp and forming a line of battle. Caesar thought they would advance to the centre of the plain to do battle, since its flatness and the fine weather provided a perfect opportunity for a decisive clash, but as the Caesareans moved forward in battle line they saw that the Pompeians had no intention of leaving their elevated position behind the boggy stream and right under the fortifications of Munda and their own camp. Despite the fact that the long march across the plain several times offered the Pompeians a favourable opportunity to attack and catch Caesar's cohorts at a disadvantage, they did not make any such move but remained under the town walls.

Caesar halted his line on his side of the stream. The position was now very dangerous because of the unfavourable ground and he therefore decided to limit his area of manoeuvre so that his whole army should not become demoralized at once by any reverse. This apparent weakness had two effects: it led Pompey to believe Caesar hesitated from fear and he began to advance his line down the hill, and it provoked Caesar's cohorts to anger because they wished to end this campaign now while the enemy was before them. The result was that Caesar's entire line crossed the marshy area and the stream and engaged the advancing Pompeians on the steep ground despite the danger of such a manoeuvre and the advantage in height it gave to the numerically superior enemy. There is no mention of Pompey's superior numbers causing his line to be longer than Caesar's so presumably the curvature of the hill restricted deployment on a wider front.

The exchange of javelins killed a vast number of men in the Pompeian ranks but when the two lines closed for close quarters fighting with the sword the struggle was so equal that for a long time the two armies remained locked together with advantage to neither side. But gradually the sheer weight of numbers and advantage of height possessed by the Pompeians began to tell and little by little Caesar's legions were forced back. His men then began to lose heart and the army was on the verge of defeat when Caesar, seizing a sword and shield, rushed into the front rank fighting and personally rallied his troops.

It was now late in the afternoon. Both sides had committed their third lines and the armies were exhausted by the stubborn, bloody fighting. At this point Caesar's favourite legion, the Xth, charged forward once more and though by now few in number succeeded in forcing the men opposite them back up the hill. This created a gap in the left wing of Pompey's line, with no reserves left to fill it. It was impossible to ignore the gradually increasing gap for the Xth, or worse still the Caesarean cavalry, which had fought the Pompeian cavalry to a standstill, might now outflank the rest of his line. In desperation Pompey withdrew a legion from his right wing and began to bring it across to reinforce his left. Seeing this manoeuvre Caesar's cavalry began attacking the weakened left wing furiously so that the withdrawn legion had to be recalled and at the same time Caesar's Numidian cavalry broke clear of Pompey's right wing cavalry and headed for the enemy's camp. Five cohorts were despatched to head off this move but the rearward movement was mistaken for a withdrawal, the whole Pompeian line was now seriously shaken by these manoeuvres which although necessary appeared to the unexperienced men in his ranks to be the beginning of defeat, and shortly afterwards the Pompeian line broke and began to fall back on Munda in confusion.

Many of the Pompeians continued to fight stubbornly during this retreat but there was now no stopping Caesar's legions and they pushed their enemies up the hill, through the camp, and into the town itself. No quarter was given and only the fortifications of the town saved those who survived the retreat.

CONCLUSIONS. Both armies were almost entirely Italian. Caesar's legions were all hardened veterans and Gnaeus Pompey too had a hard core of veterans, though many of his men were young and inexperienced. Pompey's veterans had fought against Caesar before and been pardoned: they knew what their fate would be if they were captured again. Caesar's veterans were sick of the Civil Wars and desired only to bring the wars to an end as quickly as possible. These points led both sides to fight with stubborn determination, no quarter was given by either side, and the battle was therefore one of the most bloody of the Civil Wars. It did, however, terminate the many years of civil war and earnt Caesar a perpetual dictatorship in 44 B.C.

CASUALTIES. Caesarean. 1,000 infantry and cavalry killed. 13 eagles were captured from the Pompeians. **Pompeian:** 30,000 infantry, 3,000 cavalry. Labienus and Varus were killed in the battle, Gnaeus Pompey escaped to sea but was killed later when he landed to obtain water. Sextus Pompey escaped but remained a harmless refugee while Caesar lived.

SECOND SPANISH CAMPAIGN: MUNDA

Second Spanish Campaign: Munda

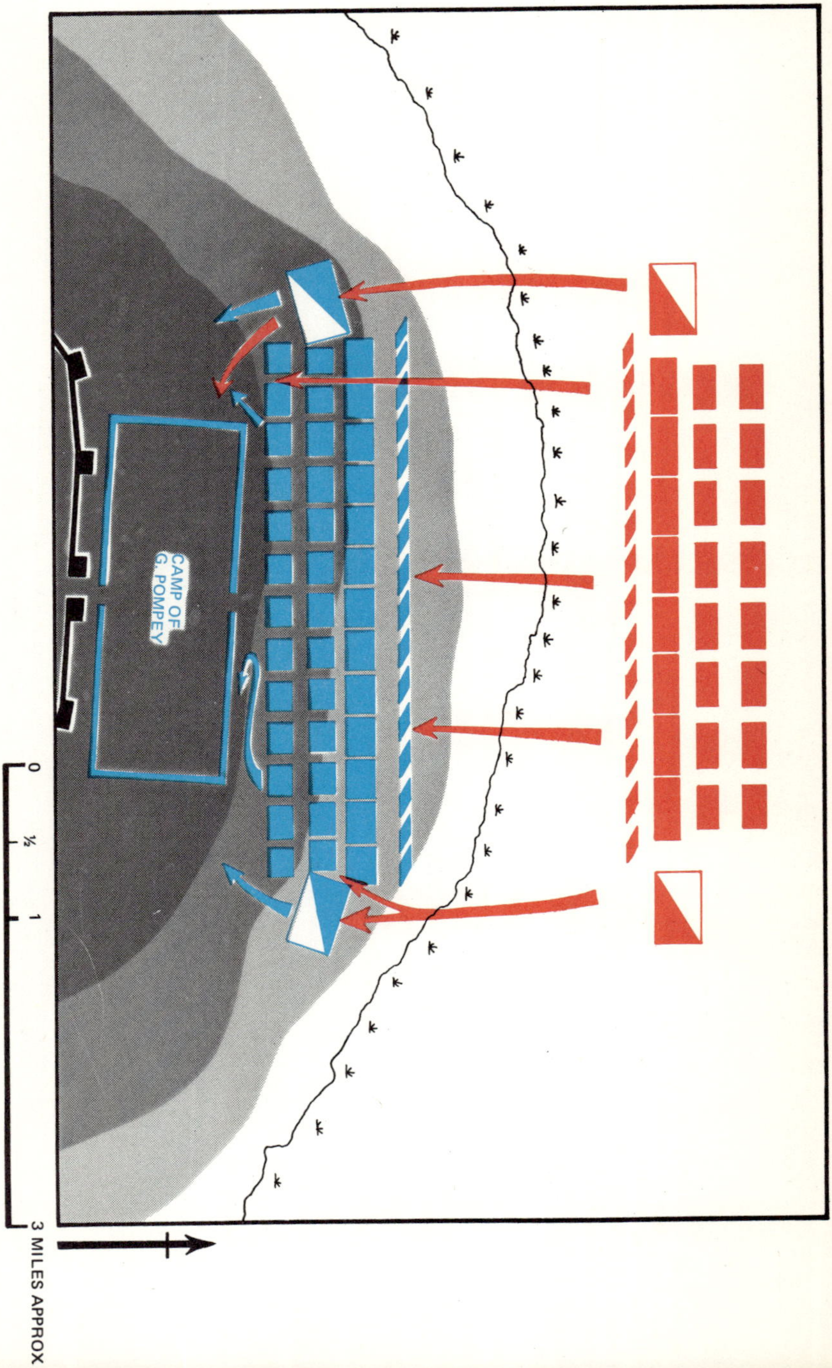

N

MUNDA The positions of the opposing forces at the beginning of the fighting, with the Caesareans about to charge across the stream to attack uphill. Note how the more numerous Pompeians have been compacted into a tighter, less pliable mass by the contours of the hill.

ORGANISATION GUIDE

The Roman legion of the 2nd Punic War had been based upon conscription and divided into hastati, principes and triarii. About 104 B.C. the consul Marius raised an army of volunteers, preferring such men to the usual assortment of reluctant small farmers conscripted against their will, who were usually of poor military value and had no interest in soldiering. This move revolutionised Roman Army organisation and from this date the Legion became a body of volunteers, many of whom were veterans – in other words professional soldiers. Instead of furnishing their own weapons, as in the civilian army, these professional soldiers were now trained and armed by the State. Consequently the differences in dress and weapons between hastati, principes and triarii, mainly caused by the finances of the individual, were replaced by standardised dress and weapons. At the same time the maniple of 120 men was replaced as the smallest tactical unit by the cohort of 600 men. A legion now consisted of 60 centuries of 80 men each, commanded by centurions, forming ten cohorts commanded by legates of the general, the first cohort being of double strength and containing the administrative services. This gave a fighting force of about 4,800 men, although the legion had a nominal strength of 6,000.

The exact organisation of the legion during the Roman Civil Wars is not known but it was little changed from the time of Marius. Weapons and armour were also practically unchanged. The legion **in the field circa** 49 B.C. was therefore probably about 4,000 men in ten cohorts with about 100 cavalry. The line of battle was now four cohorts in the front line with three cohorts in each of the second and third lines. The spaces between cohorts within a legion were gradually reduced until the lines were almost continuous. Other notable changes were the extensive use of artillery and engineers, the latter being responsible for the massive field fortifications erected during the course of the war.

The number of slingers, archers and light troops was increased considerably and the centuries of these troops were organised in the same way as the legion centuries. About 400 such troops would accompany each legion in the field but might be grouped for battle into Cohort Quingenaria (six centuries) or Cohort Milliaria (ten centuries).

The number of auxiliary cavalry was also greatly increased owing to the recruitment of mounted men from Gaul, Germany, Numidia and other nations where horsemen were predominant. Perhaps 800 auxiliary cavalry might now accompany each legion, organised in Turmae of 32 men. For battle the Turmae were grouped either as Ala Quingenaria (16 Turmae) or Ala Milliaria (24 Turmae). Both Ala and Turma fought in a two line formation.

OFFICERS

Each century in a legion had a centurion, with an optio as his second in command and a vexillarii or standard bearer. H.Q. was made up of a legatus legionis in command, six military tribunes as staff officers, an equilifer or eagle bearer, and the equivalent of the modern service and supply units – medical, engineering, clerical, orderlies, armourers etc.

MODEL FIGURES

All figures used in the photographs of the games were Airfix, except for one or two metal figures used for legati. A number of conversions were necessary:-

Arabs	–	Numidian heavy infantry and cavalry; Syrian archers.
Indians	–	Numidian light infantry and cavalry; Gallic and German auxiliary cavalry; Thracian cavalry.
Robin Hood	–	Slingers; Spanish infantry, Cretan and Spartan archers; Thessalian cavalry.
Romans	–	Roman legions; Italian and Macedonian cavalry.
Ancient Britons	–	Slingers and archers.

DRESS AND EQUIPMENT GUIDE

ROMAN CIVIL WARS

Nationality	Tunic	Cloak	Helmet	Shield	Body Armour	Weapons
Legionaries	Red, purple, green, brown, blue, yellow or unbleached with coloured borders*	As tunic	Bronze, crested	Large oval, Legion symbol	Iron mail or leather corslet	Pilum and gladius
Spanish auxiliaries Hither Spain	Unbleached, purple border	Black	None	Long, oval, wooden	None	Heavy javelin and sword
Spanish auxiliaries Further Spain	White, purple border	None	None	Light round leather target	None	Light javelin,short sword
Numidian heavy infantry	Unbleached	None	Bronze	Large, round, leather	None	Long spear and sword
Numidian light troops	Unbleached	None	None	Large, round, leather	None	Javelins, axe or dagger
Cretan Archers	Yellow, green, red or unbleached	None	None	None	None	Bow
Syrian Archers	Long robe with short sleeves, unbleached with borders	Dark colours, borders	Spiral cap	None	None	Bow
Spartan Archers	Red	None	None	None	None	Bow
Gallic cavalry	Yellow, red, purple, brown, striped or tartan	Black, striped or tartan	Bronze	None or small round	None	Spear and long sword
German cavalry	Dark colours topped by leather jerkin	Brown, red or green	Bronze, boar's tooth type	Small, light and round	None	Sword and light javelin
Numidian cavalry	Unbleached	None	None	Small, round leather	None	Javelins, axe or dagger
Macedonian cavalry	Crimson	None	Bronze, crested	Small and round	Bronze plate corslet	Short spear and kopis
Pontic cavalry	Unbleached	None	Pointed cap	Light and round	None	Javelins, sword or axe
Thracian cavalry	Black	Tartan or check	Foxskin cap	Small and round	None	Long lance and kopis
Thessalian cavalry	Yellow, red or light green	None	Bronze	None	None	2 javelins and kopis

* Unbleached wool might be cream, grey, fawn or even light brown in colour. Choice of tunic colour for the Legions was primarily left to the Legate.